D0860427

DuPONT-HABLEY

LOOKING AT HOW GENETIC TRAITS ARE INHERITED WITH GRAPHIC ORGANIZERS

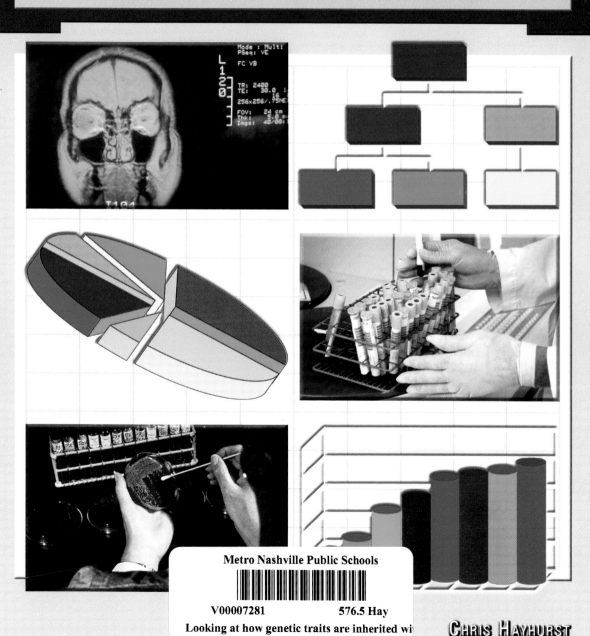

Metro Nashville Public Schools

V00007281   576.5 Hay

Looking at how genetic traits are inherited wi

CHRIS HAYHURST

The Rosen Publishing Group, Inc., New York

DuPONT-HADLEY

Published in 2006 by The Rosen Publishing Group, Inc.
29 East 21st Street, New York, NY 10010

Copyright © 2006 by The Rosen Publishing Group, Inc.

First Edition

All rights reserved. No part of this book may be reproduced in any form without permission in writing from the publisher, except by a reviewer.

Library of Congress Cataloging-in-Publication Data

Hayhurst, Chris.
Looking at how genetic traits are inherited with graphic organizers/Chris Hayhurst.—1st ed.
    p. cm.—(Using graphic organizers to study the living environment)
Includes bibliographical references.
ISBN 1-4042-0612-4 (library binding)
1. Heredity. 2. Genetics.
I. Title. II. Series.
QH438.5.H38 2006
576.5—dc22

                                                            2005018972

Manufactured in the United States of America

On the cover: A bar chart *(top right)*, a pie chart *(center)*, and a flow chart *(bottom left)*.

Contents

INTRODUCTION 4

CHAPTER ONE THE CELL 7

CHAPTER TWO CELLULAR REPRODUCTION 13

CHAPTER THREE THE GENE'S JOB 21

CHAPTER FOUR THE NATURE OF INHERITANCE 26

CHAPTER FIVE GENETICS AND THE FUTURE 33

GLOSSARY 41

FOR MORE INFORMATION 43

FOR FURTHER READING 45

BIBLIOGRAPHY 46

INDEX 47

INTRODUCTION

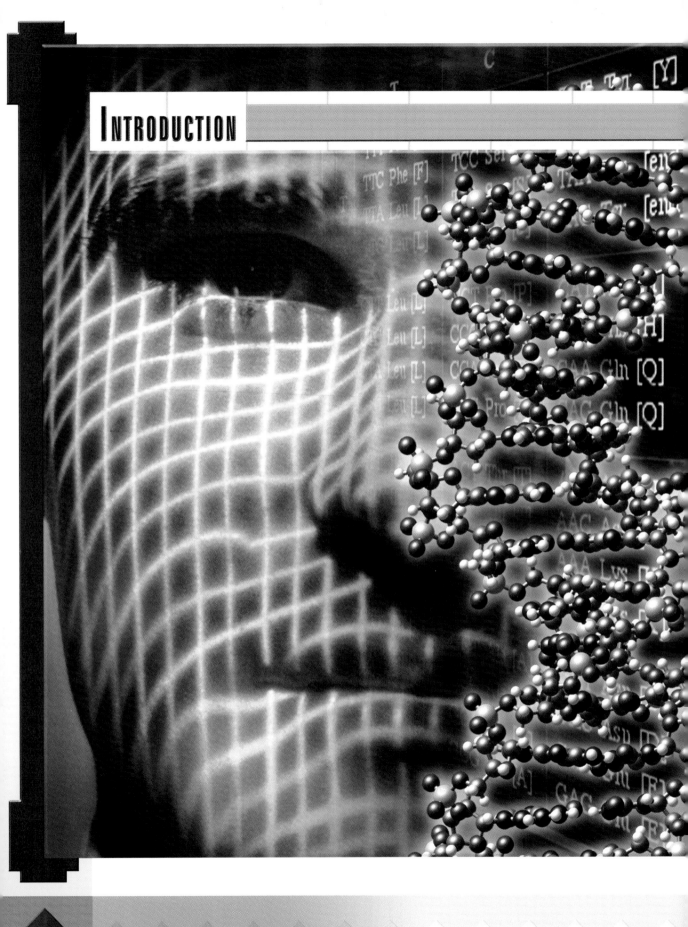

In the early 1850s, an Augustinian monk named Gregor Mendel began to study botany—the science of plants—at the University of Vienna. Like the other monks in his religious order, he lived at a monastery. The monastery, a quiet campus of thick-walled stone buildings and open, rolling fields, was set in the hills of southern Moravia, a region that today is part of Europe's Czech Republic.

Mendel was fascinated by pea plants. But even more interesting, in his opinion, was what those plants could tell him about inheritance, or the passing of traits from one generation to the next. Mendel knew that all living organisms—whether they were dogs, fish, people, or plants—reproduce. Dogs make dogs, people make people, and so on. He also knew that offspring, or the young of the species, look more like the specific individuals who made them (their parents) than other individuals of the same species. A human baby, in other words, should resemble its mother and father. Its traits, or characteristics, are similar to those of its parents. Those traits are said

Human genes are what make us different from all the other species on Earth. Whether it be elephants, humans, or geese, genes give each species its characteristics.

to be inherited. We inherit certain characteristics from our parents, who inherited their characteristics from their parents, and on and on. Many of the details of our existence—from eye color to hair thickness to height, shape, and intelligence—depend in large part upon our relatives.

For Mendel, who was far ahead of his time in his understanding of inheritance, the monastery garden was the perfect laboratory. Pea plants were easy to work with. They reproduced easily, grew fast, and displayed themselves clearly in tall, orderly columns. He could breed his plants to his heart's content, carefully combining one parent pea with another to get just the offspring he wanted. He could then take those offspring and repeat the process all over again. Today, almost 150 years later, Mendel's observations serve as the foundation for modern genetics, the scientific study of heredity.

As you might expect, genetics is quite a complicated subject. For example, in 2003, scientists had identified nearly all of the 25,000 human genes. So as you read this book, which is about the science of genetics, try to imagine Mendel in his garden, toiling away to unlock the secrets of inheritance. Also, use the included graphic organizers—charts, maps, webs, diagrams, and other visual aids—which are here to help explain difficult concepts. Each one will bring you closer to understanding what genetics is all about.

CHAPTER ONE

THE CELL

Of you're trying to get to know someone, it helps to go to their home. You can see where and how the person lives and get a feel for what he or she does. The same holds true with genetics. To understand genes—the carriers of the information that determines inheritance—you should first know where they're found, which is in the cell.

WELCOME TO THE CELL

All living organisms are made of cells. Your body—everything, including your fingernails, kneecaps, stomach, heart, and lungs—is an elaborate collection of tiny cells arranged in just the right way to make you who you are.

In nature, the simplest of organisms have just one cell. For example, the amoeba, a parasite found in water and soil and known for its ability to make people sick, has one cell. At the other end of the spectrum, complex organisms like humans have trillions of cells. Everything an organism does depends on its cells. And specific cells have specific functions. For example, if you wink at someone, special nerve cells called neurons send a command from your brain that tells your eyelid to move. Muscle cells then move your eyelid to wink.

Cells are so small they can only be seen with a microscope. Their size, though, is misleading. Even the smallest of cells are full of even smaller parts called organelles. They're called organelles because they act like miniature organs. In the same way organs of the body—such as the heart, liver, and kidneys—have specific jobs to do, the organelles in a cell have a variety of special functions. Some organelles bring in food and digest nutrients. Others rid the cell of wastes and poisons. Organelles called mitochondria produce energy that the cell needs to work. Each organelle does its part so the cell can survive.

All cells include one especially large organelle called the nucleus. The nucleus is the control center of the cell and is important to genetics because of what it contains. Inside the nucleus is a substance called chromatin. Under a high-powered microscope, chromatin looks like a cluster of tangled string.

Chromatin, which is lumped together into bigger structures called chromosomes, is made of deoxyribonucleic acid (DNA) and

CELL STRUCTURE: FUNCTION WEB

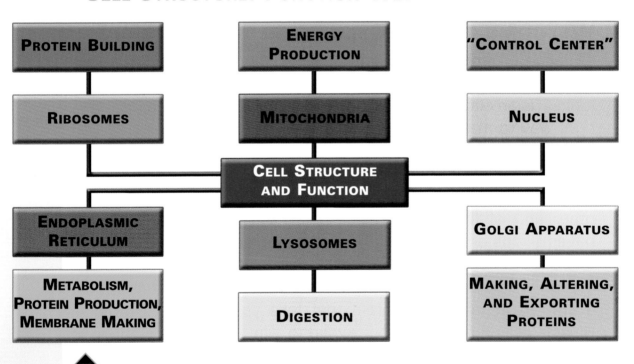

A function web is a graphic organizer that illustrates the job of each part of a system. In this case, the system is the cell. The parts of the cell are the ribosomes, mitochondria, nucleus, endoplasmic reticulum, lysosomes, and the golgi apparatus.

proteins. The protein in chromatin primarily serves a structural function, helping to maintain the chromatin's shape. DNA, on the other hand, is the key to all genetics, as we'll see.

CHROMOSOMES

The genes we inherit come packaged in what are called chromosomes. There are two main types of chromosomes: autosomes and sex chromosomes. Autosomes are the chromosomes that determine all the characteristics of a person aside from his or her sex or gender. The sex chromosomes determine only whether a person will be male or female. Males have one X chromosome and one Y chromosome (XY). Females have two X chromosomes (XX).

In human somatic (nonreproductive) cells, there are twenty-two pairs of autosomes and just one pair of sex chromosomes. In reproductive cells, the chromosomes don't come as pairs. Each cell instead has twenty-two single autosomes and a single sex chromosome.

Chromosomes are made of DNA, which is organized into genes. DNA is made of four different base units: thymine (T), adenine (A), cytosine (C), and guanine (G). Different genes consist

GENE LOCATION E-CHART

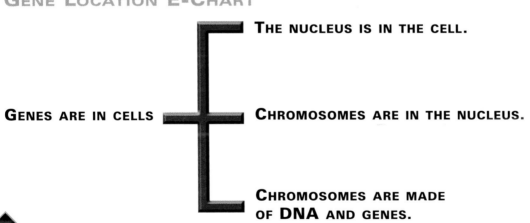

GENES ARE IN CELLS

THE NUCLEUS IS IN THE CELL.

CHROMOSOMES ARE IN THE NUCLEUS.

CHROMOSOMES ARE MADE OF **DNA** AND GENES.

In this E-chart, the main idea on the left is the general subject of where genes are located. The three ideas on the right supply additional information about where genes are located. Genes are in chromosomes. Chromosomes are in nuclei. Nuclei are in cells.

DNA LADDER

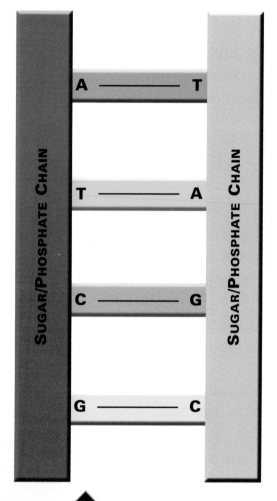

This DNA ladder clearly illustrates how DNA is structured chemically. Each rail of the ladder is composed of sugars and phosphates. These are held together by the rungs, which are made of the four nitrogenous bases.

of these four bases arranged in varying groups of three. One gene might start with TAC, for example. Another might start with AGG. An organism's genome includes all of the DNA in its cells and every individual gene.

THE DOUBLE HELIX

In 1953, American scientist James Watson and Englishman Francis Crick, working together in London, England, made a discovery that would change science forever: they determined the three-dimensional structure of DNA. Normally, figuring out how something is built is not nearly as important as knowing what that something does. In the case of DNA, however, structure is everything.

Watson and Crick found that DNA occurs in the shape of a double helix. Picture a ladder with rails on its side and rungs up the middle. A DNA molecule is a lot like that ladder, with two exceptions. First, the entire ladder is twisted like a spiral staircase, or in a helical shape. And second, its rails and rungs are actually tiny molecules, molecules so small, in fact, that they can only be seen with a high-powered microscope.

The rails of a DNA molecule are long chains of sugar and phosphate (phosphate contains the chemical phosphorous). And the "rungs" of DNA are nitrogenous bases, called such because they contain nitrogen. The nitrogenous bases are thymine, adenine, cytosine, and guanine.

Each long sugar-phosphate rail, or chain, has nitrogenous bases attached to it and sticking out from its side. The nitrogenous bases jutting out from one of the chains bond together with the nitrogenous bases of the other chain. The result is the formation of a solid "ladder." Watson and Crick determined that adenine always linked with thymine (A-T), while cytosine always linked with guanine (C-G). So an A-T linkup, for example, would constitute one complete rung of the DNA ladder.

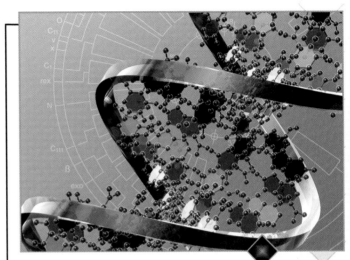

This computer rendition of the DNA double helix illustrates just how it got its name: the two, or "double," rails are arranged in a spiral "helix" form, connected by the four nitrogenous bases.

Watson and Crick built a large, three-dimensional model of DNA out of wire. The model clearly showed the double-helix, spiral-like shape of DNA, with the nitrogenous base pairs linking the spirals together. It was an amazing sight, showing the beauty of nature in its every twist and turn. Even more important, though, was what the model implied. The structure of DNA, as Watson and Crick had it, perfectly explained how the molecule replicated itself. The answer to the mystery of how genes are copied and thereby passed along to every cell in the body was staring them in the face.

DNA REPLICATION

Remember that ladder? Well here, during DNA replication, or the reproduction of DNA, is where things get interesting. In DNA replication, the ladder essentially splits down the middle, lengthwise. Imagine each rung of the ladder snapping in two, and then the two sides falling away from each other. The DNA molecule does the same thing. Signals from special proteins within the cell

DNA Replication Sequence Chart

Topic: DNA Replication

Start: Proteins inside the cell send a signal to nitrogenous bases.

Next: Nitrogenous bases break their bonds.

Next: Adenines and thymines separate. Cytosines and guanines separate.

Next: Complementary strands of bases arrive.

Next: The new bases attach to the exposed originals.

Next: Thymines link with adenines. Cytosines link with guanines.

Finish: Linkage is complete and two new DNA strands are formed.

The sequence chart presents a clear visual map of a series of events, or a sequence. This DNA replication sequence chart involves every stage of the process, from the splitting of the DNA rails to their recombination.

tell the nitrogenous bases to break their bonds. The adenines separate from the thymines. The cytosines separate from the guanines, and the two strands pull apart.

This, of course, leaves the nitrogenous bases from each strand dangling on their own. But not for long. Soon after the initial separation, new complementary strands of nitrogenous bases move in and attach themselves to the two original DNA strands. Every A on the complementary strand links with a T from the original strand. Every T links with an A. Every C links with G, and every G with C. When the linkups are complete, there are two whole "ladders" of DNA. The original single molecule of DNA is now two.

Chapter Two

Cellular Reproduction

Cells reproduce for many reasons. Sometimes new cells are required for growth. Other times, cells are needed for repairs to damaged tissue following injuries. If you scrape your shin in a fall, for example, new cells will grow to replace those that were destroyed.

All cells reproduce through a process called cell division. During cell division, a parent cell essentially splits in half, sending all of the genetic information it contains to each of its daughter cells. You might think that one daughter cell gets half the information from the parent cell while the other daughter cell gets the other half. But that's not how it works. In order to ensure that the entire genome—every last genetic message contained in a cell's DNA—is passed along to every new cell, a more complicated process is required. That process is known as mitosis.

MITOSIS

The process of mitosis, during which a parent cell divides in two to produce two daughter cells, takes place in five stages. Those stages are prophase, prometaphase, metaphase, anaphase, and telophase.

Before a cell can divide in two, however, it must first go through some changes. Picture a balloon. If you had one balloon and wanted to turn it into two balloons the same size as the original, you'd have to do a little magic. First, you'd blow the balloon up so it was twice as big as when you started. Cells

MITOSIS FLOW CHART

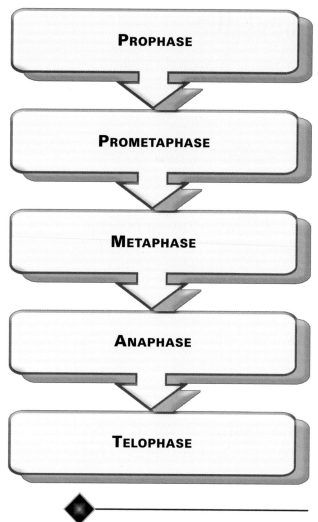

PROPHASE

PROMETAPHASE

METAPHASE

ANAPHASE

TELOPHASE

This mitosis flow chart is a good visual representation of the process to help you remember each step involved.

do essentially the same thing. To prepare for mitosis, they go through a long period of growth called interphase. During interphase, the cell copies all of its organelles, essentially doubling everything about itself. Most important, the cell's chromosomes—containing the cell's DNA and, therefore, its entire genetic package, or genome—are copied. By the time the cell has doubled everything, and interphase ends, it has nearly doubled in size.

Mitosis takes place immediately after interphase. The purpose of mitosis is to make sure that all the chromosomes (we now have twice as many as when we started) are divided evenly between the two daughter cells. This division takes place in the nucleus of the parent cell, as that's where the chromosomes are found. The first step is prophase.

PROPHASE

Two major events occur during prophase: one in the nucleus of the parent cell, and the other outside of the nucleus. In the nucleus, the threadlike strands of chromatin in each duplicated chromosome are wound into a very tight, thick bundle. The duplicated chromosomes, now called chromatids, line up in pairs.

CELL GROWTH AND DIVISION IDEA RAKE

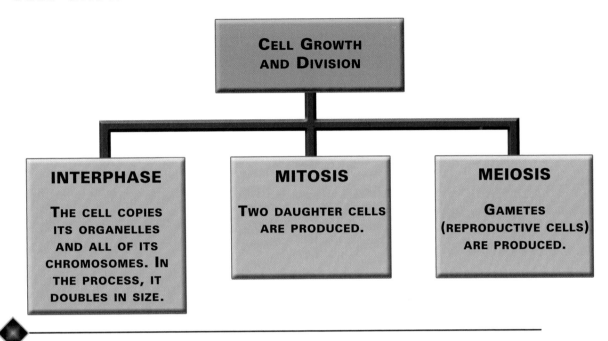

An idea rake, such as the one shown, divides a concept into its main parts. Here, the concept is cell growth and division. The stages of cell growth and division are interphase, mitosis, and meiosis.

Each chromatid is linked to its identical "sister" chromatid at its centromere—a centromere is just a narrow point near the middle of a chromatid. Imagine taking two long and narrow balloons and squeezing them together near their middles with one hand. That's what the sister chromatids look like during prophase.

Meanwhile, outside of the nucleus from either end of the parent cell, long, stringy fibers begin to climb like tentacles toward the nucleus. The fibers, made of protein, originate from two tiny, star-shaped objects called centrosomes.

PROMETAPHASE

Every nucleus has a thin, protective shell around it called a nuclear membrane. During prometaphase, the nuclear membrane surrounding the parent cell's nucleus cracks and falls apart. Some of the long protein fibers—the tentacle-like strands reaching in from the cell's centrosomes—arrive at the nucleus at the same time. Because the nucleus is no longer protected, they enter it easily. Inside the nucleus, fibers from both directions attach themselves to the centromeres of the sister chromatids.

METAPHASE

"Meta" means "middle." During metaphase, the centrosome fibers pull the sister chromatids to the exact middle of the cell—a line called the metaphase plate. By the time the tugging is over, the chromatid pairs are lined up so that each pair has one sister chromatid on either side of the metaphase plate.

ANAPHASE

During anaphase, the sister chromatids are separated from each other as the fibers pull at them from opposite directions. Now individual chromosomes, they're dragged away from the metaphase plate and toward either end of the cell. In humans, forty-six chromosomes go one way and forty-six go the other.

TELOPHASE

Telophase is the last stage of mitosis. With the duplicated chromosomes now gathered in identical groups, in pairs on either side of the cell, the cell begins to stretch. The stretching is caused by fibers from the centrosomes that bypass the nucleus during prophase and instead keep growing longer and longer until they meet similar fibers from the other side. Once they meet, these fibers push against each other. As they do the cell stretches.

As the cell stretches, the original nucleus divides in two and daughter nuclei form around the chromosome groups. At the same time, pieces from the original broken nuclear membrane mix together to form new nuclear membranes around the daughter nuclei. By the time telophase is over, the parent cell has two completely independent daughter cells within its membrane. The daughter cells are genetically identical to each other and to the original parent cell.

Now there are two daughter cells, but they're still inside the parent cell—they have yet to go out on their own. One more step has to occur for the division to be complete. That step is called cytokinesis. At cytokinesis, the parent cell, which is now very stretched out, splits in two. Cell division is over.

SEXUAL REPRODUCTION

Now that you know something about mitosis, you may be wondering: where does that first cell, the one that makes that first division to produce the first daughter cells, come from? The answer, for many organisms, lies in sexual reproduction.

When mammals sexually reproduce, the male and female mate through sexual intercourse. Soon after intercourse, the sperm from the male unites with an ovum (egg) from the female. This fertilized sperm-ovum combination is called a zygote. It's a fertilized egg, and it's held in the female's body. It's the very first cell of the new offspring.

So how do genes get passed along from the parents to the offspring? To find out, you have to look at the chromosomes.

TIC-TAC-TOE CHART

FEMALE

MALE			
	X	**X**	
X	**XX** FEMALE	**XX** FEMALE	
Y	**XY** MALE	**XY** MALE	

Like the game tic-tac-toe, a tic-tac-toe chart takes the form of a grid. In this case, it shows that combining either the X or Y chromosome from the male with the X chromosome of the female results in male or female offspring.

CHROMOSOMES

Each pair of chromosomes in an offspring's somatic cells includes one chromosome from the mother and one from the father. The chromosomes pair up according to the genes they carry. If a chromosome from the mother, for example,

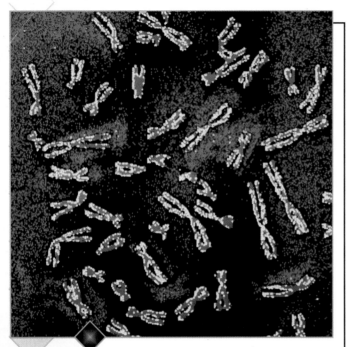

These chromosomes are visible with the aid of a thermograph, a device that uses heat to create a visible image.

carries a gene for curly hair, it will pair with a chromosome from the father that also carries a gene for hair (curly, straight, wavy, etc.). These chromosomes are called homologous chromosomes. Furthermore, those similar genes from the mother and father will be located at the same spot on each homologous chromosome. This spot is known as a locus. The female sex chromosomes (XX) are homologous. The male's (XY) are nonhomologous.

Different plants and animals have different numbers of chromosomes in their cells. In most human cells—the somatic there are forty-six chromosomes in the nucleus (two pairs of twenty-three). The only human cells that don't have forty-six chromosomes are the reproductive cells—the egg and sperm cells (the sex cells of a female and male, respectively). The reproductive cells each have twenty-three chromosomes. There's good reason for this. If the egg and sperm cells had forty-six chromosomes each, when they united at fertilization the result would be a zygote with ninety-two chromosomes. That's double what's required. Since the sperm and egg have just twenty-three chromosomes each, the result is a forty-six-chromosome zygote when they unite. Other plant and animal species maintain similar relationships between their somatic cells and their reproductive cells, which are also known as gametes. The gametes always have half the number of chromosomes as the somatic cells. In humans, reproductive cells end up with twenty-three chromosomes each through a process called meiosis.

MEIOSIS

Like mitosis, meiosis is a process of cell division. In mitosis, all forty-six chromosomes of the parent cell are passed on to each of two daughter cells. Meiosis, though, results in four daughter cells with twenty-three chromosomes each. These daughter cells are the gametes. In humans, gametes are produced—and meiosis takes place—only in the reproductive organs: ovaries in females and testes in males.

Just before meiosis takes place, the parent cell copies its chromosomes. This process, which is identical to that which

MEIOSIS TIME-ORDER CHART

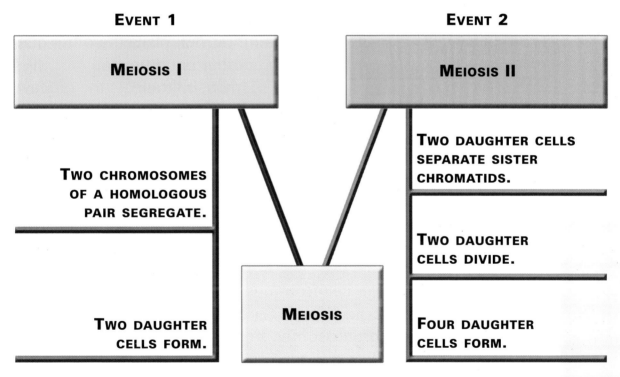

EVENT 1	EVENT 2
MEIOSIS I	**MEIOSIS II**

TWO CHROMOSOMES OF A HOMOLOGOUS PAIR SEGREGATE.

TWO DAUGHTER CELLS SEPARATE SISTER CHROMATIDS.

MEIOSIS

TWO DAUGHTER CELLS DIVIDE.

TWO DAUGHTER CELLS FORM.

FOUR DAUGHTER CELLS FORM.

This time-order chart illustrates the two main phases of meiosis: meiosis I and meiosis II. Below each phase category are the events that take place within each phase, which allows you to see the overall process.

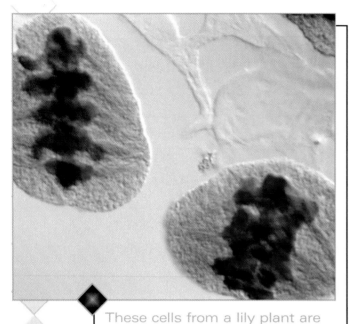

These cells from a lily plant are undergoing metaphase during meiosis I.

occurs before mitosis (interphase), produces an exact copy of all the DNA from the parent cell. Each chromosome from the parent cell becomes two sister chromatids.

Meiosis then takes place in two steps: meiosis I and meiosis II. During meiosis I, the two chromosomes in each homologous pair of the duplicated parent cell are segregated. One chromosome from each pair, each containing two sister chromatids, goes to one daughter cell, while the other goes to another daughter cell.

Meiosis II repeats the process. Without replicating their chromosomes, the two daughter cells separate their sister chromatids. Each daughter cell then divides and sends one sister chromatid from each pair to two more daughter cells. When meiosis II is complete, there are a total of four daughter cells, each with twenty-three chromosomes.

Scientists call cells with forty-six chromosomes diploid cells. Those with twenty-three chromosomes are haploid. Somatic cells, therefore, are diploid. Sex cells are haploid. If a sex cell from a male unites with a sex cell from a female and fertilization occurs, a forty-six-chromosome (diploid) zygote results. This zygote contains twenty-three chromosomes from the male and twenty-three from the female. All of the zygote's genetic information is then passed on to every new somatic cell through mitosis.

CHAPTER THREE
THE GENE'S JOB

What exactly does a gene do? How can a gene influence your body and determine your height, hair color, or even your intelligence level? The answer can be found in a long chain of events that begins with your personal collection of inherited DNA and ends with the expression of that DNA in how you look, act, and think. Scientists call that collection of DNA your genotype. The DNA's physical expression is known as your phenotype.

Remember that DNA is shaped like a double helix and is found packaged into chromosomes. Each DNA strand includes a long series of nitrogenous bases made of thymine, adenine, cytosine, and guanine. It's those nitrogenous bases that constitute the genes. Individually, T, A, C, and G have no effect on your body. But in groups of three, they control everything in your body. These groups of three are called triplet codes. There are thousands of triplet codes in your DNA, and each (like CGT, for example) contains a short message that ultimately leads

GENOTYPE VS. PHENOTYPE T-CHART

GENOTYPE	PHENOTYPE
GATACA . . .	BLUE EYES
CGTAAC . . .	RED HAIR
TTAGCG . . .	6'4" TALL
CTACTA . . .	FAIR SKIN

This T-chart pairs hypothetical genotypes with their respective phenotypes so you can see the physical results of genes.

DNA   MRNA   AMINO ACIDS   PROTEINS   **TRAIT**

This DNA-to-traits timeline illustrates the relationships between the different molecules that lead to the expression of a specific trait. Transcription produces RNA from DNA. The RNA is then translated into amino acids, which make up proteins.

TRANSCRIPTION TIC-TAC-TOE CHART

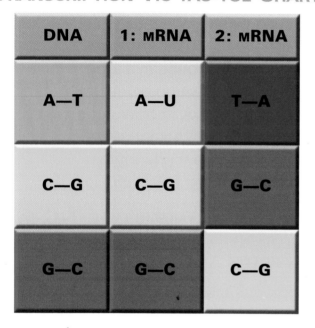

DNA	1: MRNA	2: MRNA
A—T	A—U	T—A
C—G	C—G	G—C
G—C	G—C	C—G

During transcription, the code contained in the DNA is copied to mRNA for transport out of the cell nucleus. The main difference between DNA and mRNA base-linking patterns is that in mRNA, adenine (A) bonds with uracil (U) instead of thymine (T). Only one of the two DNA strands is used in transcription. The chart above shows the two possible mRNA molecules that could result from the given DNA molecule.

to the creation of proteins, which in turn influence almost everything about you.

TRANSCRIPTION

The first step in the journey from genotype to phenotype is called transcription. During transcription, the code contained in the DNA is transcribed, or copied, to another molecule. The copying is necessary because DNA can't leave the nucleus of the cell. It's stuck there.

To solve this problem, the DNA relies on another molecule as a messenger. That molecule is called messenger ribonucleic acid (mRNA). There are three important differences between mRNA and DNA: mRNA has ribose for its sugar (not deoxyribose); it has uracil instead of thymine as one of its four nitrogenous bases; and most important, it can leave the nucleus.

Inside the nucleus, the information contained in each DNA triplet code is copied over to an mRNA molecule. Only one of the two strands of each DNA molecule is used to make the mRNA (the other strand is saved for DNA replication). In DNA replication, A links with T and C links with G. But in mRNA synthesis, a base called uracil (U), substitutes for T. So for every A on the DNA strand, a U appears on the mRNA strand. For every C, there's a G; for every G, there's a C; and for every T, there's an A. If the triplet code on DNA is AGG, the RNA will create a new nucleotide triplet of UCC. The job of mRNA is to get the transcribed message from DNA out of the nucleus and into the cell cytoplasm where it can be read.

TRANSLATION

Now that the genetic code is transcribed and out of the nucleus, it must be translated into something useful. If you were given a set of instructions on how to build a house, but those instructions were written in a language you didn't understand, you would need someone to translate the instructions to you. The mRNA contains a detailed message from DNA, but it can't understand that message, just as you can't understand the building instructions written in a language you don't understand.

Before we go any further, you should first know something about proteins. Proteins, which are relatively big molecules, are made of amino acids. There are twenty different kinds of amino acids. Different combinations of amino acids make up different proteins.

This scientist is viewing a chart of DNA sequencing, or the order of the nitrogenous bases in a certain section of DNA.

The message that mRNA contains from DNA is detailed instructions on what amino acids the cell should build. Every nucleotide triplet in the mRNA is an instruction to make one amino acid. The cell reads the triplet and translates it into something it can understand: an amino acid. Series of triplet codes for amino acids link

NATURE VERSUS NURTURE VENN DIAGRAM

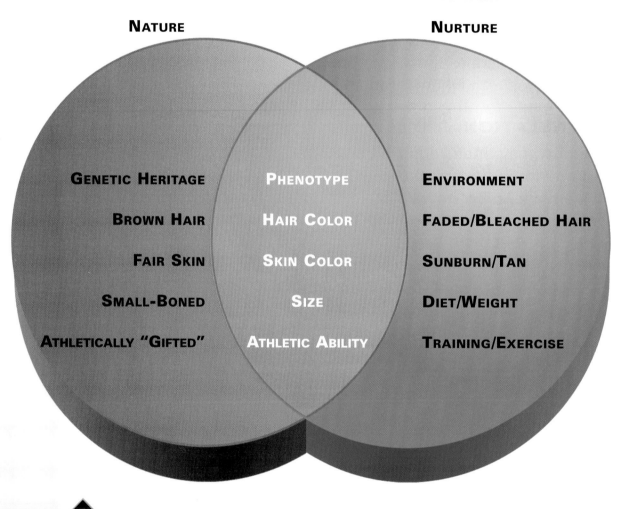

NATURE	PHENOTYPE	NURTURE
GENETIC HERITAGE		ENVIRONMENT
BROWN HAIR	HAIR COLOR	FADED/BLEACHED HAIR
FAIR SKIN	SKIN COLOR	SUNBURN/TAN
SMALL-BONED	SIZE	DIET/WEIGHT
ATHLETICALLY "GIFTED"	ATHLETIC ABILITY	TRAINING/EXERCISE

Whether nature or nurture is responsible for the behavior and appearance of organisms has been a running debate. The fact is that in certain instances, nature and nurture both contribute to the outcome of an organism. As this Venn diagram illustrates, the effects of nature and nurture overlap.

together to form thousands of different proteins. Those proteins, in turn, each have a specific job to do, such as to create brown hair.

ENVIRONMENTAL INFLUENCES

You've probably heard the phrase "nature versus nurture." People often wonder whether we're shaped by our genetic heritage (nature) or by the way we're brought up and the influences of the environment around us (nurture). People will argue one side or the other. They'll maintain that everything about a person can be traced back to their gene pool. Or

This Cornell University researcher is studying how rice grows when it's influenced by genetic modification.

they'll say that where a person lives, where they go to school, the friends they meet throughout life, and the world around them are what shape them.

The truth is, both nature and nurture play major roles in who we are and even how we look. Thanks to genes from your mother, you may be born with brown hair, but years later, that hair might fade to blond as you spend time in the sun. Meanwhile, as you tan during the summer, your skin may grow temporarily darker than the tone you were born with. You may be genetically inclined to grow really tall, but if you don't get enough to eat as a child and your diet is nutritionally deficient, you may not grow to your full genetically determined height. And even if both of your parents are world-famous track stars, among the fastest people on the planet, your genetic gift will likely remain hidden if you have no interest in sports and never put on a pair of running shoes. Who you are depends on so many factors, not just your genes.

Chapter Four

The Nature of Inheritance

You might wonder why you resemble your parents in some ways but you don't look exactly like them. This feature of inheritance is due to something called genetic recombination. Because an organism's inherited chromosomes come from both its mother and its father, it won't be genetically identical to either. Instead, it will have some genes from one parent and some genes from the other. Two siblings with the same parents will usually also be genetically different, both from each other and from their parents. Every offspring is born with a unique combination of genes pulled from their parents' genome.

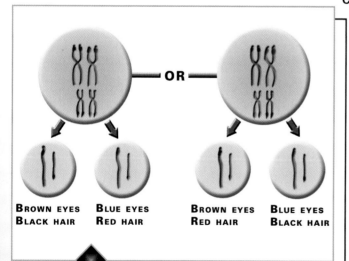

BROWN EYES BLACK HAIR **BLUE EYES RED HAIR** **BROWN EYES RED HAIR** **BLUE EYES BLACK HAIR**

As this illustration shows, chromosomes can line up more than one way during meiosis, which can result in different phenotypes for the offspring.

A MATTER OF CHANCE

When homologous pairs of chromosomes line up along the midline of the parent cell during meiosis, they do so randomly. Imagine asking twenty-three couples, each consisting of one man and one woman, to hold hands. You then tell them to all line up on a gym floor along the same painted line, and ask each person to keep the line between them and their partner. You'd have twenty-three pairs of people, still holding hands, with the line between each pair. Now, look at the result. Some couples are facing one way, while some face the other. Depending on how each couple first held hands, some men will be on the right side of the line, while some will be on the left.

When chromosomes line up, the same thing happens. The maternal chromosome (from the mother) of each pair has a 50-50 chance of being on one side of the midline or the other. The paternal chromosome (from the father) has the same odds. Therefore, when the parent cell splits and the daughter cells are formed, which chromosomes go where depends on how they lined up. Some paternal chromosomes will go one way with some maternal chromosomes, while other maternal and paternal chromosomes will go the other direction. The result is that the daughter cells (gametes) each have a random combination of maternal and paternal chromosomes. This process is known as independent assortment and is one means of genetic recombination. Mendel

GENETIC RECOMBINATION E-CHART

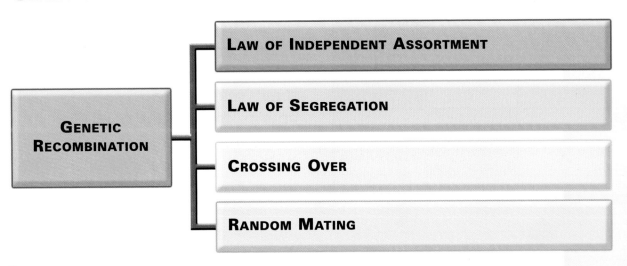

This E-chart shows the component parts of genetic recombination. Each of these parts contributes to genetic recombination.

discovered the law of independent assortment using the pea plants in his garden.

THE LAW OF SEGREGATION

Mendel made other discoveries in his garden as well. One had to do with how specific characteristics are passed from a parent somatic cell to its daughter gametes during meiosis. Homologous chromosomes pair up according to the genes they carry. So a chromosome from the mother that carries her gene for eye color will pair up with a chromosome from the father that carries his gene for eye color. These two different genes controlling the same characteristic are called alleles. Mendel's law of segregation says that alleles separate from each other during meiosis. What this means for the gamete is simple: it receives the alleles that come with its chromosomes.

ALLELE TIC-TAC-TOE CHART

	A	A
A	AA HOMOZYGOUS DOMINANT	AA HETEROZYGOUS
A	AA HETEROZYGOUS	AA HOMOZYGOUS RECESSIVE

This allele tic-tac-toe chart pairs up dominant and recessive alleles into three possible combinations.

DOMINANT VERSUS RECESSIVE

When organisms sexually reproduce, their gametes (reproductive cells) combine at fertilization into a zygote, the new offspring. When they meet, each gamete (the sperm from the male and the ovum from the female) brings with it all of its chromosomes and all of the genes found in those chromosomes. In humans, remember, each haploid gamete contains twenty-three chromosomes.

The characteristics that are passed on to the zygote depend on the alleles carried by each

THE LIFE CYCLE

chromosome. But not all of those alleles will be expressed in the organism's appearance. If, for example, the sperm brings an allele for blond hair while the ovum brings an allele for black hair, only one of those alleles will become evident when the child grows hair. Which one it is depends on the alleles themselves. When a so-called dominant allele on one chromosome is paired with a recessive allele on its homologous chromosome, the dominant allele will be expressed. A recessive allele will be expressed only if its homologous chromosome also carries that recessive allele.

The creation of life is a cycle, as this diagram shows. From fertilization to mitosis to meiosis, one organism grows and creates another.

An organism is said to be homozygous for a specific trait when it has two identical alleles (recessive or dominant) in a homologous pair of chromosomes. An organism is heterozygous for a trait when its homologous chromosomes carry two alleles that are different.

OTHER SOURCES OF GENETIC VARIATION

Another way organisms pass genetic information is through a process known as crossing over. Crossing over takes place during meiosis. When homologous chromosomes line up in pairs, they're so close together—attached, in fact—that they sometimes trade parts. When this happens, genes from one chromosome wind up on the other chromosome, and vice versa. The result is that every chromosome that goes to the gamete has at least a gene or two from both the mother and the father.

Genetic variation also occurs because sexually reproducing organisms can, in theory, mate with any other organism of the opposite sex within the same species. People, for example,

GENETIC DISORDER COMPARE AND CONTRAST CHART

GENETIC DISORDER	
DOMINANTLY INHERITED	**RECESSIVELY INHERITED**
CAUSED BY DOMINANT ALLELES	CAUSED BY RECESSIVE ALLELES
RR AND Rr (HOMOZYGOUS OR HETEROZYGOUS)	RR (HOMOZYGOUS RECESSIVE = SHOWS DISORDER) Rr (CARRIER OF RECESSIVE ALLELE = DOES NOT SHOW)
HUNTINGTON'S DISEASE ACHONDROPLASIA (DWARFISM)	CYSTIC FIBROSIS SICKLE-CELL ANEMIA

As this compare and contrast chart illustrates, genetic disorders are either dominantly inherited or recessively inherited. By comparing the two columns, you can see the differences between dominantly and recessively inherited genetic disorders.

reproduce with other people from entirely different parts of the world. When two people mate, they each bring their entire genetic heritage with them. Then, depending on which alleles made it to the sperm and ovum that meet at fertilization, a fraction of that heritage is passed on to their offspring.

GENETIC DISORDERS

Some genetic disorders occur when genes fail to make proteins with important functions. Others occur when those proteins are made, but are not made correctly or don't work like they should. Genetic disorders are

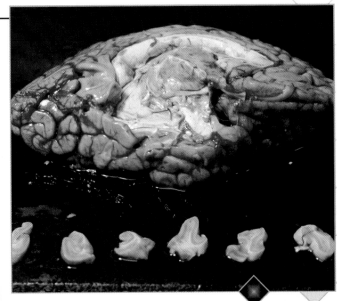

This human brain is from the Harvard Brain Tissue Research Center, which studies brain tissue from people who died of genetic disorders.

classified as either recessively inherited or dominantly inherited. Dominantly inherited disorders are the result of dominant alleles of genes. Recessively inherited disorders are due to recessive alleles.

People with recessively inherited disorders inherit a recessive allele from each parent. They're said to be homozygous for that allele. Cystic fibrosis, a genetic disease that leads to deadly infections, is one example of a recessively inherited disorder.

Some people carry the recessive allele for genetic disorders but never show signs of those disorders. These people are heterozygous for the allele. They also carry a dominant, healthy form of the allele, so the recessive allele has no effect on them. Because they carry the recessive allele, however, they can potentially, and unknowingly, pass the disorder to their children if their partners are also heterozygous for the disease.

Dominantly inherited disorders occur when the dominant allele is the cause of the disease. People with dominantly inherited disorders can be either homozygous or heterozygous for the problem allele. Huntington's disease, which affects the nervous system, is one example of a dominantly inherited disorder.

MUTATIONS

Sometimes mistakes occur during DNA replication, causing changes to the genes. When such a mistake occurs in a somatic cell, all future cells deriving from that cell will also carry the mistake. Mutations are typically not dangerous unless they affect the construction or function of an important protein. The only way a mutated gene can be passed along to offspring is if the change occurs in a reproductive cell.

CHAPTER FIVE

GENETICS AND THE FUTURE

Mendel laid the foundation for the study of genetics. Before him, little was known about genes or heredity. After him, the field's potential was essentially limitless. Mendel presented the findings from his studies at a meeting of a local scientific society in 1865. There, he never mentioned the word "genes." "Genetics" was not in his vocabulary, nor was it in the vocabulary of anyone else in attendance. What he talked about, though, had everything to do with genetics. His peas had everything to do with the future.

The word "gene" was eventually coined nearly thirty-five years after Mendel's presentation. Scientists finally had a name for the microscopic units of information that their predecessor had discovered were the key to inheritance. Genetics, the study of heredity, was born.

This researcher is studying a map of the human Y chromosome for the U.S. Human Genome Project. At the time of this photograph, the Y chromosome was the only chromosome to have been fully mapped.

MODERN GENETICS

The science of genetics has come a long way since Mendel's day. Today, new developments in the field make news headlines all the time. In the summer of

GENETIC BREAKTHROUGHS WHEEL

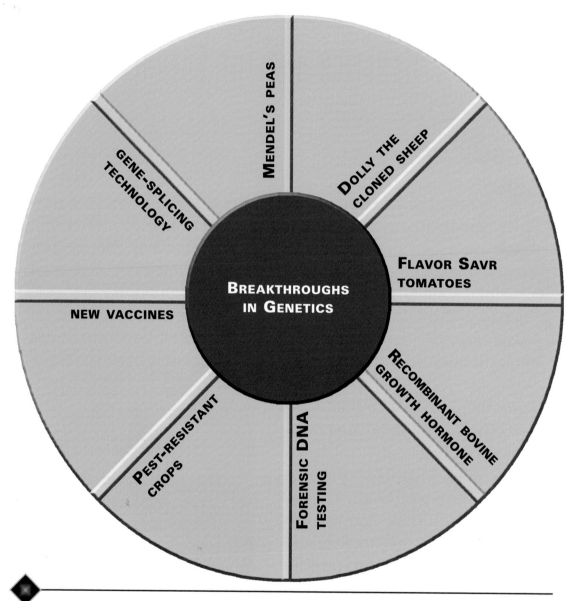

A wheel organizer illustrates a central idea, in this case breakthroughs in genetics, around which are its components. Here the components are various discoveries and technologies. A lot of developments have come out of genetics, such as new vaccines and DNA testing.

1996, it was Dolly the sheep, the first cloned mammal. Scientists had inserted the DNA of an adult sheep into another sheep's egg. That egg, with its new DNA, then began to divide like a normal fertilized egg. The embryo grew, and Dolly, a perfect copy of the DNA supplier, was born. Due to both biological problems and ethical issues that cloning raises, human clones have never been made. But some predict they will be in the future.

Now that scientists at the U.S. Human Genome Project have achieved their goal of identifying all of the nearly 25,000 genes

FACT AND OPINION CHART

GENETIC ENGINEERING	
FACT	**OPINION**
GENETICALLY ENGINEERED (GE) FOODS ARE SOLD ON SUPERMARKET SHELVES AND ARE NOT ALWAYS LABELED AS SUCH.	GE FOODS SHOULD BE LABELED SO I CAN AVOID THEM.
	I DON'T CARE IF GE INGREDIENTS ARE IN MY FOOD OR WHETHER I EVEN KNOW IT.
GE CROPS ARE GROWN IN THE UNITED STATES.	GE CROPS SHOULD NOT BE PERMITTED ANYWHERE IN THE UNITED STATES.
	THE UNITED STATES SHOULD CONTINUE TO BE A LEADER IN THE CULTIVATION OF GE CROPS.
TRANSGENIC ANIMALS ARE USED IN MEDICINE TO STUDY DISEASES.	THE USE OF TRANSGENIC ANIMALS TO STUDY HUMAN DISEASES IS UNFAIR TO THOSE ANIMALS AND INHUMANE.
	SUCH STUDIES ARE NECESSARY AND PERFECTLY ACCEPTABLE IF IT MEANS HUMAN LIVES WILL BE SAVED.

Genetic engineering is a controversial issue. This is why fact and opinion charts are useful. Here we have two columns. The fact column provides an unbiased fact. The opinion column offers contrasting viewpoints on the fact with which they're associated.

found in human DNA, work continues as geneticists study those individual genes and learn each one's importance to the human body. In other developments, gene therapy is being explored as a way to fix the genes of those born with genetic disorders. DNA testing is used in courtrooms to identify criminals and set innocent people free. A debate is emerging about whether sports professionals should be allowed to add specific genes to their DNA to improve the athletic abilities of their children.

GENETIC ENGINEERING

Genetic engineering is a laboratory technique in which scientists take certain genes from one organism and give them to another.

This researcher for the genetic engineering company Monsanto is extracting a corn embryo from a genetically modified plant.

Genetic engineers work to invent desirable gene combinations that would otherwise never exist. One of the first genetically engineered foods available at supermarkets was the Flavr Savr tomato, which appeared in 1994. The tomato, which looked normal—if not a little too perfect—included a gene from a flounder that increased its resistance to frost and prevented it from spoiling. Other products of genetic engineering include recombinant bovine growth hormone (rBGH), which is used in cows to help them produce more milk. There's even a special soybean designed to survive repeated applications of its manufacturer's own herbicide, allowing farmers to spray their crop without fear of destroying it.

CAREERS IN GENETICS CLUSTER WEB

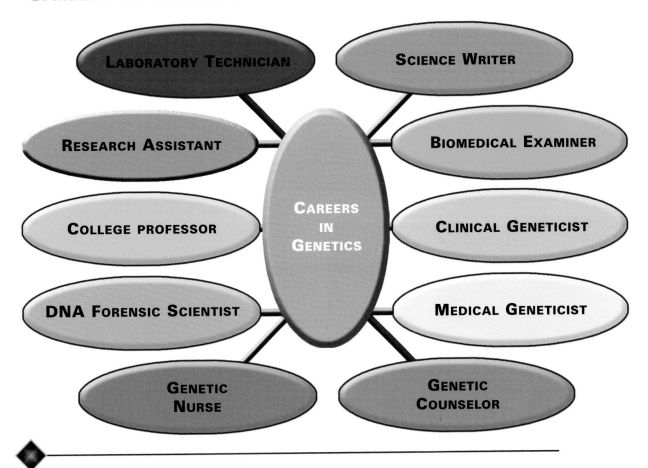

Genetics is a large, dynamic, and diverse field. In this cluster web, you can see a sampling of just a few of the careers available to people in the field of genetics.

In medicine, genetic engineering has allowed scientists to alter animals so they develop human diseases. Scientists can then study those diseases and learn how to treat them in people. Some think a cure for cancer may be found through genetic engineering. Other researchers are hard at work looking for genes that lead to long and healthy lives. Those genes could be a modern-day fountain of youth. Many other developments are also in the works.

THE GREAT DEBATE

Environmentalists argue that products of genetic engineering could lead farmers to use unnecessary chemicals to control weeds. They could also lead to more toxic chemicals in the water and food supply. Organic farmers fear accidental cross-pollination between genetically engineered crops and their own non-genetically engineered crops. Currently there is no evidence that any of the genetically engineered products on sale in the United States are harmful to consumers. But some believe that testing procedures used to determine their safety are not good enough.

Genetic engineering of food is more prevalent than many people think. These common groceries all have genetically modified ingredients.

Genetically engineered foods also have their supporters, of course. And many of the same things that environmentalists see as problems are considered improvements by those in the industry. For instance, the fact that it's possible to insert a gene into a plant to make the plant resistant to a particular chemical allows farmers to kill weeds without killing their crops. Other benefits, they say, can already be seen on store shelves: low-fat french fries from potatoes engineered to require less oil for frying, rice with more essential nutrients, and fruits and vegetables with greater amounts of vitamins E and C. Other products include peppers with improved taste and color, and firmer, more colorful tomatoes. Those who argue for genetically engineered foods believe these products and others are the world's answer to a growing population faced with less and less farmland.

The debate over the safety and environmental integrity of genetically engineered foods will likely continue for a long time. For now, though, at least two-thirds of all foods offered in the United States contain ingredients from genetically engineered crops. Genetically engineered products, it seems, are everywhere.

MARCHING ON

There's an old saying: Don't blink, as life may pass you by. The same can be said for genetics. Turn away for even a few seconds and you're likely to miss something. Genetics research is

GENETICS INVERTED TRIANGLE

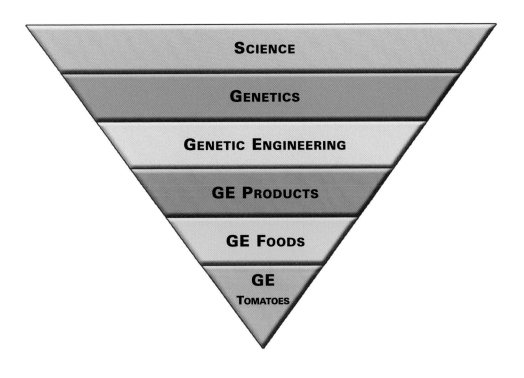

SCIENCE

GENETICS

GENETIC ENGINEERING

GE PRODUCTS

GE FOODS

GE TOMATOES

The inverted triangle illustrates the degree of specificity of a subject. Here in the genetics inverted triangle, the general subject is science. As you move down the pyramid, the subjects become more and more focused.

making great strides almost every day. Things are moving so fast, in fact, that scientists themselves can hardly keep up with new developments.

What genetics research will bring us in the future is anybody's guess. But one thing is for sure: the more that is known about genes, heredity, and life in general, the more life itself will change. For anyone interested in genetics, now is an exciting time. What's happening today is one thing. What happens tomorrow may be beyond your wildest dreams.

allele One of two forms of a gene, each of which produces variations of the same biological characteristic.

autosome A chromosome that determines characteristics of a person aside from his or her sex.

centrosome A star-shaped object in a cell responsible for producing fibers during cell division.

chromatid A duplicated chromosome formed during DNA replication.

chromatin A DNA-protein mixture of chromosomes.

chromosome A structure carrying genetic information that is found in the cell nucleus.

cytokinesis The last step in cell division, resulting in two daughter cells.

diploid cell A cell with a set of chromosomes inherited from each of two parents.

DNA Deoxyribonucleic acid, or a molecule carrying genetic information found in chromosomes.

embryo An organism at any stage prior to birth.

gametes Reproductive cells (sperm and egg cells) that unite during sexual reproduction.

gene The microscopic building block of inheritance made of DNA nucleotides.

genetic recombination A process by which offspring inherit combined characteristics of both parents.

genetics The scientific study of genes and heredity.

genome The entire collection of genes found in an organism.

genotype An organism's genetic constitution.

haploid cell A cell containing one set of chromosomes (twenty-three chromosomes for humans).

heterozygous Having two different alleles of a gene for one characteristic.

homologous chromosomes Chromosomes that pair up because of genetic similarities.

homozygous chromosomes Chromosomes having identical alleles of a gene for one characteristic.

inheritance The passing of traits from one generation to the next.

locus A specific gene's exact location on a chromosome.

meiosis A process of cell division in which sexually reproducing organisms form reproductive cells, or gametes.

mitosis A five-phase process of cell division during which a parent cell divides in two to produce two daughter cells.

nuclear membrane A thin, protective covering surrounding a cell's nucleus.

nucleus A large organelle containing a cell's chromosomes.

phenotype An organism's physical appearance.

RNA Ribonucleic acid, or a molecule that carries genetic information provided by DNA from a cell's nucleus to be translated into the amino acids and proteins that make a cell work.

sexual reproduction A process in which the gametes from two parents unite to form a fertilized egg.

transcription A process during which the code contained in the DNA is copied to form RNA.

translation A process during which the genetic "message" carried by mRNA is used to make amino acids and proteins.

triplet codes Three-nucleotide groups that determine what amino acids are formed during translation.

zygote A fertilized egg cell.

FOR MORE INFORMATION

American Museum of Natural History
The Gene Scene
Central Park West at 79th Street
New York, NY 10024-5192
(212) 769-5400
Web site: http://ology.amnh.org/genetics

Genetic Science Learning Center at the Eccles Institute
  of Human Genetics
University of Utah
15 North 2030 East, Room 2160
Salt Lake City, UT 84112-5330
(801) 585-3470
Web site: http://gslc.genetics.utah.edu

Genetics Education Unit
Murdoch Children's Research Institute
10th Floor, Main Building
Royal Children's Hospital
Flemington Road
Parkville, Victoria 3052
Australia
+6 (138) 341 6298
+6 (138) 341 6212
Web site: http://www.genecrc.org/site/ko/index_ko.htm

National Human Genome Research Institute
National Institutes of Health
Building 31, Room 4B09
31 Center Drive, MSC 2152
9000 Rockville Pike
Bethesda, MD 20892-2152
(301) 402-0911
Web site: http: //www.genome.gov

WEB SITES

Due to the changing nature of Internet links, the Rosen Publishing Group, Inc., has developed an online list of Web sites related to the subject of this book. This site is updated regularly. Please use this link to access the list:

http://www.rosenlinks.com/ugosle/gtgo

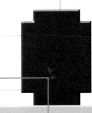

FOR FURTHER READING

Bailey, N. C., and N. L. Eskeland. *Call Me Gene*. 2nd ed. Del Mar, CA: Science2Discover, 2001.

Bailey, N. C., and N. L. Eskeland. *My Name Is Gene*. 2nd ed. Del Mar, CA: Science2Discover, 2002.

Beatty, Richard. *Genetics*. Austin, TX: Raintree Steck-Vaughn, 2001.

Butterfield, Moira. *Genetics: Present Knowledge, Future Trends*. London, England: Franklin Watts, 2002.

Gonick, Larry, and Mark Wheelis. *The Cartoon Guide to Genetics*. New York, NY: HarperPerennial, 1991.

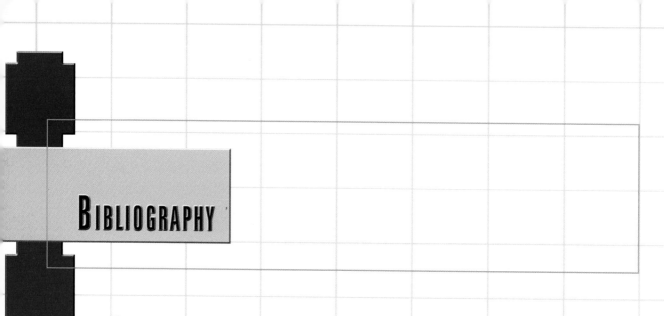

BIBLIOGRAPHY

Boron, Walter F., and Emile L. Boulpaep. *Medical Physiology: A Cellular and Molecular Approach*. Philadelphia, PA: Elsevier Science, 2003.

Campbell, Neil, Jane Reece, and Lawrence Mitchell. *Biology*. 5th ed. Menlo Park, CA: Benjamin-Cummings Publishing, 1999.

Ganong, William F. *Review of Medical Physiology*. 19th ed. Stamford, CT: Appleton & Lange, 1999.

Henig, Robin Marantz. *The Monk in the Garden*. New York, NY: Houghton Mifflin Company, 2000.

Karp, Gerald. *Cell and Molecular Biology*. 2nd ed. New York, NY: John Wiley & Sons, 1999.

A

adenine, 9, 11, 12, 21
alleles, 28–29, 31
amino acids, 23, 24

C

cell
  diploid, 20
  haploid, 20, 28
  overview of, 7–8
  reproductive, 9, 18, 32
  somatic, 9, 18, 20, 28, 32
cell division
  meiosis, 18, 19–20, 26, 28, 29
  mitosis, 13–14, 16, 19, 20
  stages of, 13–17
chromatin, 8–9
chromosomes, 9, 21
  autosomes, 9
  in cell division, 14, 16, 19, 20,
    26–27, 28–29
  overview of, 17–18
  sex chromosomes, 9
cloning, 35
Crick, Francis, 10, 11
crossing over, 29–30
cystic fibrosis, 31
cytosine, 9, 11, 12, 21

D

deoxyribonucleic acid (DNA)
  nitrogenous bases of, 10–11, 12,
    21, 22
  replication of, 11–12, 32
  structure of, 10–11, 21
  transcription of, 22–23
  translation of, 23–25
  triplet codes, 21–22, 23, 24
Dolly, 34–35

G

gametes, 18, 19, 27, 28
gene therapy, 36
genetic disorders, 31–32, 36
genetic engineering, 36–37
  debate on, 38–39
genetic recombination, 26, 27
genetic variation, 29–31
genome, 10, 14, 26, 35
genotype, 21, 22
guanine, 9, 11, 12, 21

H

Human Genome Project, 35–36
Huntington's disease, 32

I

inheritance, 5–6

M

Mendel, Gregor, 5–6, 27–28, 32
messenger ribonucleic acid
   (mRNA), 22, 23, 24
mutation, 32

N

nature versus nurture, 25

P

phenotype, 21, 22
proteins, 8, 11, 22, 23,
   24–25, 32

R

ribonucleic acid (RNA), 23

S

segregation, law of, 28
sexual reproduction, 17, 28, 30

T

thymine, 9, 11, 12, 21, 22

U

uracil, 22, 23

W

Watson, James, 10, 11

ABOUT THE AUTHOR

Chris Hayhurst has an MS degree in human anatomy and physiology. He is also the author of numerous young adult books on biology and nature, including *Animal Testing: The Animal Rights Debate*, *The Brain and Spinal Cord*, *Learning How We Move*, *Cool Careers Without College for Animal Lovers*, *Everything You Need to Know About Food Additives*, *Everything You Need to Know About Hepatitis C*, and *The Lungs: Learning How We Breathe*.

PHOTO CREDITS

Cover, pp. 1, 5 (top, middle, and bottom) © Medical Library Volume 2/ Royalty-Free/Gazelle Technologies, Inc.; pp. 4–5 © William Whitehurst/ Corbis; pp. 8, 9, 10, 12, 14, 15, 17, 19, 21, 22 (top and bottom), 24, 26, 27, 28, 29, 30, 34, 35, 37, 39 courtesy of Nelson Sá; p. 11 © Corbis; p. 18 © Howard Sochurek/Corbis; p. 20 © Clouds Hill Imaging Ltd./Corbis; p. 23 © Andrew Brookes/Corbis; pp. 25, 36 © Jim Richardson/Corbis; p. 31 © Rick Friedman/Corbis; p. 33 © Roger Ressmeyer/Corbis; p. 38 © Touhig Sion/Corbis.

Designer: Nelson Sá; Editor: Nicholas Croce; Photo Researcher: Nelson Sá